Field Guide to Snow Crystals

Edward R. LaChapelle

International Glaciological Society   Cambridge

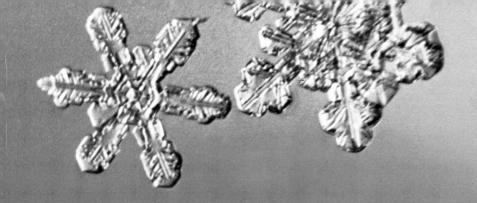

FIELD GUIDE

TO SNOW CRYSTALS

Preface

This book began in 1952 when I built a portable camera to photo-graph snow crystals during a study of the compaction of snow into glacier ice on the Juneau Ice Field in Southeastern Alaska. This study was part of a glaciology research project sponsored by the American Geographical Society. In 1953 I took the camera with me to Green-land for similar studies on the Greenland Ice Cap. During these two field seasons I was simply using photography to record the size and shape of particles found in glacier snow and firn. A systematic study of snow crystals was quite the furthest thing from my mind.

At the same time that these glacier studies were in progress, I had begun working in the winters on avalanche research for the United States Forest Service at Alta, Utah, a job which I have continued each winter to the present time. By 1956 it had become apparent that the very large natural variation in snow crystal forms played an important but poorly understood role in avalanche formation. In that year I began to photograph snow crystals during the winter storms, noting the character of weather and avalanches each time a picture was taken. The 1952 glacier crystal camera, which had been gathering dust for three years, was dug out of storage and put to work. By 1960 this record of snow crystals had acquired sufficient importance that a formal study of their relationship to avalanche formation was incorporated into the regular program of the Alta Avalanche Study Center. Additional camera equipment was adapted to this work and an improvised cold laboratory set up. By 1966 I was able to present a paper at an international snow conference in Japan describing some results of this study.

In the meantime, a large number of snow crystal photographs had accumulated. The value of this collection began to appear when I found myself involved in training programs for snow rangers as the Forest Service expanded its snow safety and avalanche observation network. Long practice had enabled me to identify a wide variety of crystals and often to describe their origin and history. Trainees found the same crystals puzzling. The photographs proved useful in lectures as a means of illustrating the different crystal types and evolution. The skeleton of this book thus was born as a collection of pictures to show snow rangers what snow crystals were really like.

There are two very well-known books already available on snow crystals. The reader interested in the subject is urged to consult both of them. One is *Snow Crystals* by W. A. Bentley (New York: Dover T287, 1962, originally published by McGraw-Hill, New York, 1931). This book is a magnificent artistic collection of photographs. It has long been the standard source for artistic illustrations of snow crystals. Unfortunately, it does not provide any scientific information about the conditions under which the crystals were collected and photographed. This latter deficit does not occur in *Snow Crystals: Natural and Artificial,* by Ukichiro Nakaya (Cambridge, Mass.: Harvard University Press, 1954). Nakaya's book has become the scientific classic on snow crystals. It also offers many beautiful photographs, together with a thorough technical discussion of their origin and character. Nakaya was able to make in the laboratory almost all the types of crystals found in nature. This gave him an opportunity to measure with great precision the atmospheric conditions which controlled their formation.

The present book is specifically intended to fill a gap left by these two well-known volumes. Both treat at great length the details of individual crystals carefully arranged under a microscope. Neither of them presents the picture that a casual observer sees when he scoops up a handful of snow and examines it with his naked eye or a small hand lens. What such an observer sees often resembles very little what he finds in the Bentley and Nakaya books. The snow ranger, the nature lover or the serious student needs a field guide to snow crystals which explains what he actually sees and how the snow got that way. This book tries to fill that need.

EDWARD R. LaCHAPELLE

Seattle, Washington

Preface to the Reprint Edition 1992

Since 1988, when the original publisher of this Field Guide allowed it to go out of print, a continued demand has encouraged the publishing of this reprint. As a basic guide for the practical field worker and as a reference for instruction, it has stood the test of time remarkably well. The original purpose of its publication, to explain the nature and origin of snow crystals as they are observed in the everday, real world, remains as valid today as it was in 1969. Although understanding of some of the details has improved over the past two decades, the broad base of scientific principles underlying the evolution of crystals in the winter snow cover remains valid. Both the scientist and the practical person in the field will therefore find this reprint is still a useful Field Guide.

The principal changes since 1969 that the reader and student of snow should note are those in nomenclature. There are two basic paths of snow crystal metamorphism in the deposited snow cover, described and illustrated in the text. One path occurs with little or no bulk temperature gradients in the snow, the other in the presence of strong temperature gradients. A brief review of the evolving terms that describe these two paths will help bring the reader up to date in the matter of nomenclature.

In 1952, the Swiss scientist Eugster, taking a morphological approach to snow crystal evolution, introduced the terms *destructive* and *constructive metamorphism* to describe the two basic paths. These terms are mentioned here in the Field Guide. At the same time that this Field Guide was originally being compiled, Sommerfeld and LaChapelle submitted a scientific paper, published in 1970, that proposed a classification of metaporphosed snow based on a genetic point of view. They suggested that the two basic paths should be called *equitemperature* (ET) and *temperature gradient* (TG) *metamorphism*, corresponding, respectively, to Eugster's destructive and constructive metamorphism. These genetic terms are the ones used throughout this Field Guide. The convenient expressions, "ET" and "TG" have since become widely used by English-speaking snow workers and continue in use to this day. Subsequent to publication of the original Field Guide, improved scientific insights have been gained into snow crystal evolution, principally through the work of Colbeck at the Cold Regions Research and Engineering Laboratory. Today, the preferred and scientifically correct terms of crystals produced by the two basic metamorphic paths are *equilibrium* and *kinetic growth forms.*

Current scientific thinking on snow crystal terminology is summarized in the "International Classification of Seasonal Snow on the Ground", published in 1990 by the International Commission of Snow and Ice (IAHS). This short reference work contains some excellent additional illustrations of crystal types as well as a detailed table setting forth terms and symbols for snow classification from both the morphological and genetic viewpoints. The reader is referred to this publication for a useful supplement and update to this reprint edition of the Field Guide to Snow Crystals.

EDWARD R. LaCHAPELLE

McCarthy, Alaska
December 1991

Contents

Field Guide to Snow Crystals

The Classification of Snow

Snow consists of ice crystals in the atmosphere which grow large enough to fall and reach the ground. There are two fundamentally different types of snow: that which falls from the sky and that produced by later changes within deposited snow. All snow crystals start as the first kind, but sooner or later they lose their original identity and become the second kind. Each has its own distinct classification scheme, the first much more complicated than the second. When an observer looks at a sample of snow, the first distinction he must make is whether he is examining *precipitated* snow or *metamorphosed** snow. Any attempts to fit one type into a category of the other type can only lead to confusion. The commonest error in this respect is to try to fit metamorphosed snow into the International Snow Classification (described below), which is intended for solid precipitation.

In addition to these two types of snow, there are the nonprecipitated forms which do not originate as falling snow at all. They include rime and hoarfrost, which are formed at the point of deposition and properly speaking are not snow at all but another type of ice crystal. These sometimes are mistaken for snow.

Precipitated Snow

All precipitation starts as water vapor in the atmosphere. Whenever a parcel of air is sufficiently cooled, some of this water vapor

---

* *Metamorphism* is a term borrowed from geology. It means change of structure under heat and pressure. The term *metamorphosis,* which has wider meanings, is not applied to snow.

condenses to form clouds. The cloud particles may be water droplets (water cloud) or they may be ice crystals (ice cloud). Because most clouds exist at higher levels in the atmosphere where the air is cold, most precipitation begins as ice particles, though much of it melts before it reaches the earth's surface. It is also rather common to find supercooled water clouds, where the water droplets still remain liquid although the air temperature is below freezing. Under certain conditions the cloud droplets coalesce or the ice crystals grow until they reach a size where they begin to fall. The processes which cause clouds to discharge precipitation are complex. Discussion of them is beyond the scope of this book, but the interested reader may wish to consult the suggested references. It will suffice here to point out that snow crystals begin as minute ice particles which have formed around condensation nuclei in the atmosphere. These nuclei may be dust particles with favorable molecular structure or even minute crystals of sea salt. If these ice particles exist in an atmosphere with an excess supply of water vapor (supersaturated air) they will continue to grow. When they reach a critical size, they begin to fall. Depending on the conditions they encounter in different layers of air during their descent, they may continue to grow, may stop growing, or may experience a succession of growth patterns. What arrives at the ground can be a simple product of the initial ice crystal formation, or it may be the result of a complex life history. The latter is especially true if the growing snow crystal is caught in an updraft and recirculated through the cloud.

If the growing snow crystals fall through a cloud composed of supercooled water droplets, they will collect a coating of *rime*. The droplets are able to remain liquid at subfreezing temperatures only as long as they remain suspended in the air. As soon as they touch any solid object, including snow crystals, they immediately freeze. Supercooled cloud droplets driven by the wind against solid objects on the ground will freeze to form the familiar accretions of rime (see Fig. 52). They form similar accretions on individual snow crystals. Much of the liquid water content of clouds which reaches the ground is extracted in the form of rime on falling snow. (Except in winter, this snow melts during its fall and reaches the ground as rain.) Evaporation from supercooled cloud droplets is also an important source of vapor for the initial growth of snow crystals.

Each water molecule consists of one oxygen and two hydrogen atoms. The arrangement of water molecules is only semistructured in the liquid. When water freezes, these molecules assume an orderly arrangement with fixed positions for the oxygen atoms. The hydrogen atoms provide the bonds which hold together this structure called the *crystal lattice*. The arrangement of this lattice in ice is such that it generates solid forms with hexagonal symmetry in one plane. This *basal* plane, which is fixed in respect to the lattice, is called the plane of the crystallographic a-axes. There are three such a-axes, each separated by 60° from the next in this plane. At right angle to this plane of hexagonal symmetry is the crystallographic c-axis, also called the principal axis of the crystal. There is no hexagonal symmetry in a c-plane. These crystallographic features and their relation to the major types of snow crystals are illustrated in Figure 1. The relative rates of growth along the different crystal axes determine which type will be formed.

At this point it is important to distinguish what the term "snow crystal" really means. A single crystal is any single ice particle which has a common orientation of the orderly array of molecules which makes up its solid structure. It may or may not exhibit external evidence of this order, such as facets, symmetry, or regular form. A regular, intricate, stellar snow crystal and a rounded particle of ice are both single crystals if they have a single *crystallographic orientation* throughout. A snow *grain,* on the other hand, is a mechanically separate particle in the snow cover. Each grain may be a single crystal, or vice versa, but this is not necessarily so. A single grain may consist of several distinct crystals, or a single crystal may embrace several cohering grains. In the types of snow discussed in this book, grain and crystal commonly coincide. The term "snow

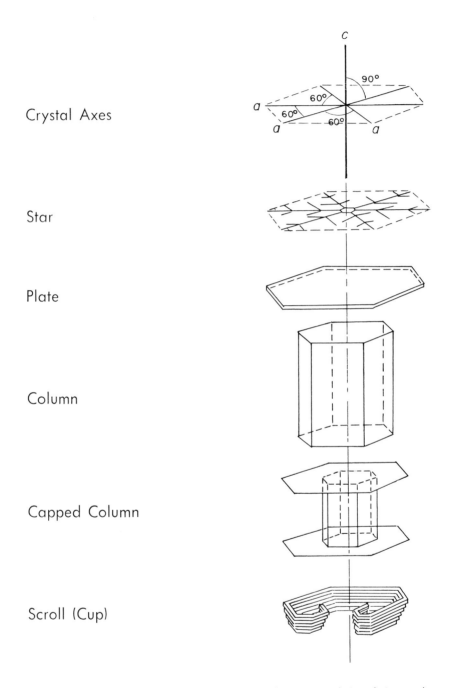

Crystal Axes

Star

Plate

Column

Capped Column

Scroll (Cup)

1. Structural arrangement of the principal types of snow crystals in relation to the crystal axes of ice. (After an illustration by Swiss Federal Institute for Snow and Avalanche Research)

crystal" will be used for all single-crystal grains, whether or not they exhibit external evidence of crystalline structure.

The atmospheric factors determining the growth rates are the temperature and the amount of water vapor available (degree of supersaturation). These factors have been examined in the laboratory by Ukichiro Nakaya, who was able to grow under controlled conditions most of the principal snow crystals types found in nature and to measure accurately the conditions under which they grew. The results of his many experiments extending over several years are summarized in the now-famous Nakaya diagram (Fig. 2). This diagram shows a series of curves plotted on a graph of supersaturation (with respect to ice) versus temperature. These curves define the boundaries of supersaturation-temperature regions within which different types of snow crystals are formed. For instance, needle-shaped crystals are seen to form only in a narrow range of temperatures around -5 to -8°C, but over a wide range of supersaturation.

The Nakaya diagram is based entirely on laboratory measurements. It has been widely assumed that this diagram approximately describes the conditions of natural snow formation in the atmosphere. More recent studies by other Japanese scientists have confirmed the essential truth of this assumption, although there have been some modifications in detail. In 1966, C. Magono and C. W. Lee published a diagram similar to that of Nakaya which was based on field measurements of atmospheric temperature and humidity during snowstorms. This diagram (Fig. 3) is similar to Nakaya's in basic features, but extends the observations to lower temperatures and includes the domain of rime formation on snow crystals. Because the actual supersaturation at subfreezing temperatures is difficult to measure in clouds, this latter feature of the Magono and Lee diagram lacks precision. The diagram does offer the advantage of having been based on direct observations in the field. For this reason it is a more reliable indicator of conditions prevailing at the cloud levels where the snow crystals are formed.

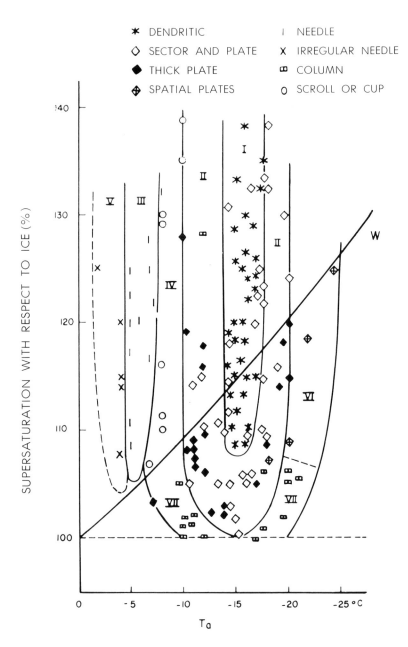

2. *Air temperature–supersaturation diagram showing the conditions of formation of various forms of snow crystals. This is the well-known "Nakaya diagram" based on observations in the laboratory. W is a line giving the saturation vapor pressure in respect to supercooled water. (After Nakaya, 1945; reproduced by permission of Harvard University Press)*

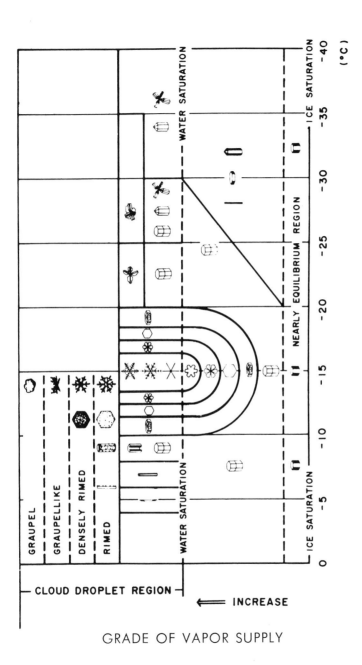

3. *Temperature and humidity conditions for formation of natural snow crystals in the atmosphere. These conditions are similar to those shown in the Nakaya diagram, but are based on actual observations in snow-forming clouds. (After Magono and Lee, 1966; reproduced by permission of the authors)*

Perfectly symmetrical snow crystals are in the minority. The thousands of regular stars and plates which make up the bulk of Bentley's collection of snow photographs are deceptive in this respect for they are the product of careful selection over many years. In fact, most snow crystals are irregular, often unsymmetric, and are modified by complicated growth regimes. Frequently they are coated with rime, although the importance of this feature varies with climate. In addition to these irregularities, the falling crystal is exposed to other vicissitudes as it falls. It may strike other crystals and become attached to them. Turbulent winds can break off branches of the fragile stellar crystals, and then these branches proceed to grow or collect rime on their own. Even after it reaches the ground, the original crystal or its fragments is still not free from modification. The wind can cause further fragmentation by drifting, or it may deposit rime by driving supercooled cloud droplets against the snow surface. When at last the snow crystal or its remnants comes to rest under an observer's microscope, it can bear little resemblance to the original particle formed high in the atmosphere. Even after reference to a detailed classification scheme, it is sometimes difficult to identify the character and origin of snow freshly deposited on the ground.

The most widely used classification for solid precipitation is that proposed in 1951 by the International Commission on Snow and Ice. A pictorial abstract of this classification scheme is presented in Figure 4. This is a simplified scheme which embraces the main structural types of snow crystals, plus graupel (heavily rimed crystals), ice pellets, and hail. The seven main crystal types are plates, stellar crystals, columns, needles, spatial dendrites, capped columns, and irregular forms. In the broad sense, most snow crystals found in nature will fit into one of these categories. There are many variations within the categories which are not separately identified by the International Snow Classification. In some instances combination types are difficult to classify. Having a category like "irregular crystals" is like having a file drawer labeled "miscellaneous." It is

Graphic Symbol	Examples			Symbol	Type of Particle
				F1	Plate
				F2	Stellar crystal
				F3	Column
				F4	Needle
				F5	Spatial dendrite
				F6	Capped column
				F7	Irregular crystal
				F8	Graupel
				F9	Ice pellet
				F0	Hail

4. A pictorial summary of the International Snow Classification for solid precipitation. This classification applies to falling snow.

	N1a Elementary needle		C1f Hollow column		P2b Stellar crystal with sectorlike ends	
	N1b Bundle of elementary needles		C1g Solid thick plate		P2c Dendritic crystal with plates at ends	
	N1c Elementary sheath		C1h Thick plate of skeleton form		P2d Dendritic crystal with sectorlike ends	
	N1d Bundle of elementary sheaths		C1i Scroll		P2e Plate with simple extensions	
	N1e Long solid column		C2a Combination of bullets		P2f Plate with sectorlike extensions	
	N2a Combination of needles		C2b Combination of columns		P2g Plate with dendritic extensions	
	N2b Combination of sheaths		P1a Hexagonal plate		P3a Two-branched crystal	
	N2c Combination of long solid columns		P1b Crystal with Sectorlike branches		P3b Three-branched crystal	
	C1a Pyramid		P1c Crystal with broad branches		P3c Four-branched crystal	
	C1b Cup		P1d Stellar crystal		P4a Broad branch crystal with 12 branches	
	C1c Solid bullet		P1e Ordinary dendritic crystal		P4b Dendritic crystal with 12 branches	
	C1d Hollow bullet		P1f Fernlike crystal		P5 Malformed crystal	
	C1e Solid column		P2a Stellar crystal with plates at ends		P6a Plate with spatial plates	

5. The meteorological classification of snow crystals according to the scheme of Magono

	P6b Plate with spatial dendrites		CP3d Plate with scrolls at ends		R3c Graupel-like snow with nonrimed extensions
	P6c Stellar crystal with spatial plates		S1 Side planes		R4a Hexagonal graupel
	P6d Stellar crystal with spatial dendrites		S2 Scalelike side planes		R4b Lump graupel
	P7a Radiating assemblage of plates		S3 Combination of side planes, bullets, and columns		R4c Conelike graupel
	P7b Radiating assemblage of dendrites		R1a Rimed needle crystal		I1 Ice particle
	CP1a Column with plates		R1b Rimed columnar crystal		I2 Rimed particle
	CP1b Column with dendrites		R1c Rimed plate or sector		I3a Broken branch
	CP1c Multiple capped column		R1d Rimed stellar crystal		I3b Rimed broken branch
	CP2a Bullet with plates		R2a Densely rimed plate or sector		I4 Miscellaneous
	CP2b Bullet with dendrites		R2b Densely rimed stellar crystal		G1 Minute column
					G2 Germ of skeleton form
	CP3a Stellar crystal with needles		R2c Stellar crystal with rimed spatial branches		G3 Minute hexagonal plate
	CP3b Stellar crystal with columns		R3a Graupel-like snow of hexagonal type		G4 Minute stellar crystal
	CP3c Stellar crystal with scrolls at ends		R3b Graupel-like snow of lump type		G5 Minute assemblage of plates
					G6 Irregular germ

and Lee. This scheme permits much more detailed classification than the International one. It also applies to falling snow.

a repository for everything that will not fit elsewhere. Especially for the casual or untrained observer, too many snow crystals turn out to be "irregular." If snow records are being accumulated for scientific or technical purposes, as is done for avalanche warning networks, valuable details are lost. Sometimes subtle distinctions among crystal types will reflect widely different conditions in the atmosphere or behavior of the deposited snow.

Nakaya developed a much more detailed classification which went a long way to eliminate the imprecision in the International Classification. Nakaya's scheme has since been modified and improved by the same Magono and Lee mentioned earlier. The Magono-Lee classification, published in 1966, is the best one currently available. It probably will accommodate 99 per cent of snow crystals commonly observed in nature. This scheme is presented in Figure 5. The carefully drawn sketches are the same as those which appear in the Magono-Lee diagram in Figure 3. Cross-checking between these two illustrations will help in understanding of the different crystal types and their origin. This classification offers some very useful distinctions among different combination types and recognizes the various degrees of riming which constitute such a common feature in many climates. Needle crystals are subdivided into needles, sheaths, and long solid columns, a useful distinction because of the widely different temperature at which these types form. (Sheaths are hollow crystals with a needlelike shape.) Another very useful subdivision is among the different types of irregular cluster (S1, S2, S3). Probably the greatest improvement of all is the inclusion of categories for broken branches (stellar fragments), for these occur very frequently. (The complete International Classification includes mention of broken crystals, but omits them from the pictorial scheme.) Magono and Lee also added an entirely new category, that of the "germ" crystal. These are snow crystals in the first stage of formation, before they have grown to "normal" size. These sometimes reach the ground if they are formed in shallow cloud layers close to the earth's surface. Note that although Magono and Lee have not entirely escaped the "Miscellaneous" trap, few crystals will be left to fall into this category.

Further details about the character and formation of individual crystal types are discussed in the text accompanying the crystal photographs.

Metamorphosed Snow

From the viewpoint of thermodynamics, newly formed snow crystals have very unstable shapes. This is especially true of the intricately branched crystals. A large ratio of surface area to volume means that the surface molecules of the crystal have a large amount of potential energy stemming from intermolecular attraction (large surface free energy). The natural tendency of thermodynamic processes (in this case the redistribution of mass and energy through phase changes) is to reduce surface free energy to a minimum. This means that snow crystals will tend to change so that the ratio of surface area to volume will approach a minimum.

The ideal shape to achieve this minimum is a sphere. Thus even the most intricate snow crystals tend in time to become rounded particles of ice. This process is called *equitemperature metamorphism,* for it proceeds only in bodies of snow which are not far from a uniform temperature. Because it tends to destroy the original shape of snow crystals, it is also known as *destructive metamorphism.* It is most rapid at temperatures close to the freezing point and diminishes in intensity as the temperature falls. Below about -40°C it virtually comes to a halt. There is considerable experimental evidence to support the idea that most of the shape changes take place by transfer of water molecules from one part of a crystal to another through the vapor phase (evaporation and subsequent redeposition as a solid). Other processes may also be at work, but this is the dominant one.

The stages of such metamorphism in a stellar crystal are sketched in Figure 6. Note that the branches of the crystal tend to break up during metamorphism, so that the initial product is a number of

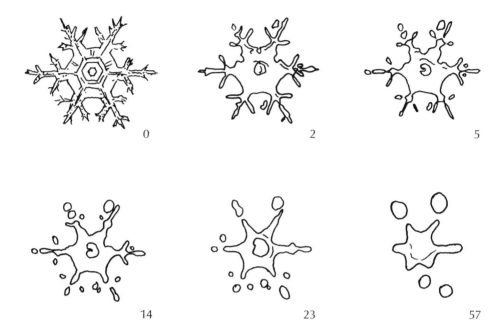

6. *The destructive metamorphism of a stellar snow crystal. The numerals give the age of the snow crystal in days.*

small ice grains rather than a single large one with the same mass as the original crystal. This gives the fine-grained appearance to *old snow,* which by definition is the end product of equitemperature metamorphism. Reduction in the space each snow crystal occupies causes the snow cover to shrink, or settle. This settlement is the normal accompaniment of metamorphism and is partially responsible for the fact that the snow accumulated on the ground is always much less thick than the total thickness of all the individual layers at the time of their deposition. Lowering of the snow surface through settlement (but not melt) gives external evidence of the metamorphic process taking place within the snow cover.

Equitemperature, or destructive, metamorphism is complete when the snow particles have been reduced to rounded grains (ideally to spheres) of ice. Such snow normally has a density of between 500 and 600 kilograms per cubic meter.* The bulk density of randomly nested uniform ice spheres is about 580 kilograms per cubic meter. Changes that lead to further density increases are called *firnification*. This can take place by two processes. One is the refreezing of meltwater trapped among the grains by capillary action. The other is compaction of the snow matrix under pressure. Both processes are normally at work in the accumulation zones of temperate glaciers. (In polar glaciers which never melt, compaction alone does the job.)

Refreezing is the dominant mechanism in the early stages of firnification, when the layers of old snow are close to the surface and subjected to winter freezing. Later, compaction becomes more important as the firn is buried under successive annual accumulations. Both processes bond the individual grains more closely together, increasing the mechanical strength as well as the density of the firn. The final product, when the communicating pores are closed off to form separate air bubbles, is glacier ice.

This normal process of metamorphism is completely altered when large differences in temperature exist between adjacent layers of snow. This steep temperature gradient imposes differences in equilibrium water vapor pressure in the interstitial spaces of adjacent snow layers. Vapor then flows from regions of higher pressure (warmer snow) to those with lower vapor pressure (colder snow). Because the vapor is composed of the same water molecules as the solid ice framework, it is not necessary for these molecules to wend their way through the intricate passages among the snow crystals. A "hand-to-hand" transfer process is at work instead. The vapor is deposited as ice on an adjacent and colder crystal, while ice on the other side of the same crystal is removed as vapor and proceeds to the next crystal.

---

* All densities given in this book follow the modern usage of the International Scientific Metric units. A density of 500 kilograms per cubic meter is equivalent to 0.5 gram per cubic centimeter, or 50 per cent of the density of water.

As the snow thus passes through the vapor stage, it is being subjected to *temperature-gradient metamorphism*. If this process, also called *constructive metamorphism,* continues, eventually the entire mass of the snow will pass through the vapor stage and be redeposited as new crystals. These crystals have an entirely different character from the original precipitated snow and its subsequent modifications. The tendency of equitemperature metamorphism to form rounded ice grains is reversed, and more complex crystals develop. If the process persists, the new crystals continue to grow and may eventually achieve a much larger size than the original precipitated crystals.

The process of temperature-gradient metamorphism is variable, depending on such factors as altitude and snow structure, but it generally prevails when temperature gradients exceed about 0.1°C per centimeter. There is a tendency for larger crystals to form with larger temperature gradients. The regeneration process proceeds most rapidly when the vapor pressure gradient is largest. In addition, the vapor pressure gradient is larger for a given temperature gradient at warmer temperatures than it is at colder ones. As a result of heat from the earth, snow is normally warmer close to the ground than at the surface; hence the changes are most rapid at the bottom of the snow cover.

The effects of such metamorphism on mechanical properties of snow are profound. The newly formed crystals are much more weakly bonded together than are those produced by equitemperature metamorphism. The latter process is accompanied by *sintering,* or the development of intercrystalline bonds. Temperature gradients produce very little sintering and result in larger crystals with fewer bonds to be sintered. The net result is weakening of the snow. Tensile strength, resistance to shearing, and load-bearing capacity all diminish drastically. If the process is carried to completion, the snow develops a very fragile structure which will collapse into a cohesionless mass of crystals on slight disturbance. This condition is sometimes called "sugar snow." The scientifically correct term is *depth hoar,* reflecting the use of *hoar* as a generic term for all kinds of ice crystals that form by sublimation directly from the vapor (see discussion below). Depth hoar has many undesirable practical effects on snow. It undermines compacted roads or runways, interferes with packing of ski runs, and is a major cause of avalanches. Sometimes its formation is carried to the point where the lowermost layers of a shallow snow cover are eroded by net evaporation and hollow spaces actually form under the snow.

Depth hoar forms most readily in cold, continental climates with shallow snow covers such as the Colorado Rockies, for these conditions favor strong temperature gradients. It is also most likely to occur early in the winter while the snow is still thin. Deep snowfalls and warm temperatures in a coastal maritime climate are much less likely to produce the necessary temperature gradients.

The visible effects of regeneration first appear as angular and faceted crystals among those that retain their original form or have previously been rounded and simplified by destructive metamorphism. As the sublimation and diffusion of water vapor proceeds, the new crystals begin to take shape and the old ones gradually disappear. The new ones assume various shapes—cups, scrolls, columns—according to the local conditions affecting their growth, but they all have one feature in common that distinguishes them as depth hoar. All such forms produced by sublimation are characterized by a layered structure that appears externally as a stepped or ribbed surface on certain crystal facets. This is the most reliable identifying feature of depth hoar. There is some tendency for the developing depth hoar crystals to align themselves in faint columnar patterns parallel to the direction of vapor diffusion.

Classification of metamorphosed snow (Fig. 7) presents a problem different from that of precipitated snow. In one way the problem is simpler, for there are fewer distinctly different forms to consider. The crystals of precipitated snow are the end products of different growth regimes in the atmosphere, but the various types of metamorphosed snow are simply stages in ongoing processes.

I. Unmetamorphosed (New) Snow	II. Equitemperature (Destructive) Metamorphism
(See Magono-Lee Classification for details)	II-A-1. Original crystal forms easily distinguishable
I-A. Little or no wind, crystals largely intact	
I-B. Wind-drift, crystals fragmented	II-A-2. Original forms distinguishable with difficulty
	II-B-1. Original forms fragmented and no longer recognizable; fine-grained old snow
	II-B-2. Rounded ice grains

grain size diminishes

grain size grows

7. *The classification of metamorphosed snow according to the scheme proposed*

III. Temperature-Gradient (Constructive) Metamorphism	IV. Firnification
III-A-1. Angular crystals, none layered (begins in new snow)	IV-A. Melt-freeze metamorphism; grains bonded by freezing
III-A-2. Small and poorly formed layered crystals	IV-B. Pressure metamorphism; grains bonded by compression and recrystallization (freezing also possible)
III-A-3. Mature, fine- or medium-grained depth hoar, prominent layering	(Glacier ice—noncommunicating pores)
III-B-1, III-B-2. Similar sequence to III-A, but begins in old snow and leads to coarse-grained depth hoar	

by Sommerfeld and LaChapelle (1969). This scheme applies to deposited snow on the ground.

Classification thus has to take into account stages of evolution rather than end products. Glacier ice is the only true end product of snow metamorphism, and even glacier crystals continue to evolve. Much winter snow terminates with complete equitemperature metamorphism only because the snow is destroyed by melt before firnification can proceed. Depth hoar can exist as an end product of temperature-gradient metamorphism only as long as the necessary temperature gradient persists. Once this is removed, the process reverts to equitemperature metamorphism. The appearance of the various classification categories cannot be fixed with precision, and the division between them is somewhat arbitrary.

Nonprecipitated Crystals

The types of snow described up to this point have either originated in the atmosphere or have been produced by subsequent metamorphism of such crystals. Two other types, which strictly speaking are not snow at all, are formed at the earth's surface. These are rime and hoarfrost. The two are readily confused in nature by an observer and even more readily confused in the literature by authors. Rime is often called hoarfrost and vice versa. The two types of deposited ice crystals have entirely different origins. Once the manner in which they form is clearly understood, it is easy to distinguish one from the other by physical appearance (see Figs. 54-58 and 61-64).

Rime has already been mentioned as a deposit of frozen water droplets on snow crystals which may also form on terrestial objects exposed to wind-driven supercooled clouds. The amount of rime so deposited may vary all the way from a light frosting to the enormous accumulations several meters thick which are found on the summits of mountain peaks constantly exposed to winter storms. In each case the rime is built up by freezing of supercooled cloud droplets as they strike the exposed surface. Depending on wind and weather conditions, rime accretions may take the form of a dense, compact mass; a feathery surface; or even slender, needlelike spikes. Close examination will always reveal a fine-grained, granular structure without any recognizable crystalline pattern. Rime always grows into the prevailing wind; it is a reliable indicator of storm wind direction. The thickness of the deposit for a given storm is also roughly proportional to the average wind velocity for that storm.

Hoarfrost is formed by sublimation of solid ice crystals directly from water vapor in the air; it is the ice equivalent of dew. This process has already been mentioned in connection with temperature-gradient metamorphism. It takes place at the snow or other surfaces whenever these surfaces are cooled below the dew-point temperature of subfreezing air. The commonest cause of such cooling is loss of radiant heat on clear, cold nights; hence hoarfrost is most commonly found following such a night when the humidity is high. Hoarfrost crystals have a definite external structure which usually exhibits the stepped or layered characteristics of crystals formed from vapor. The smooth facets of hoarfrost crystals glitter in sunlight, quite in contrast to the dull, matte surface of rime. A coating of such crystals on the snow surface is called *surface hoar*.

Snow Crystal Observation and Photography

Many snow crystals can easily be recognized with the naked eye, but the dedicated observer will soon find that a modest amount of magnification will help him to examine the smaller forms. A simple pocket magnifier is very useful. These come in a variety of sizes, some offering a combination of individual lenses with options for different magnifying powers. The compact kind easily carried in the pocket will usually be the most useful in the long run, for it is more likely to be taken into the field as a matter of habit. A 10X Coddington magnifier serves well.

Much more detail in individual crystals can be seen with greater magnification, such as 25X to 40X. This requires a microscope, a firm place to set it, illumination, and a cold room of some sort in which to work. Low-power binocular microscopes are very advantageous, for they are easily focused and offer a large working distance between specimen and microscope objective lens. Quite satisfactory instruments can be obtained for as little as one hundred dollars, but the very fine microscopes are far more expensive. Illumination without heat is essential; a small flashlight will serve very well as an improvised light source. The manner of arranging the illumination is discussed below in connection with snow crystal photography.

There are two methods of photographing small objects. One is called *photomicrography,* the other *photomacrography.* Photomicrography adapts a camera body or a film holder to some version of a standard microscope. The optical system forming the image on the film consists of the lenses of the microscope. This is the only convenient optical method for taking pictures of very small objects. Photomacrography utilizes both the body and the lens of standard camera equipment with the addition of lens extension tubes or supplementary lenses to permit focus at very short distances. The image-forming system makes use of the camera lens itself. This latter system is best suited to photographing objects ranging in size from a millimeter to a few centimeters in diameter.

Because camera lenses have much larger apertures than microscope lenses, exposures times are usually much shorter for photomacrography than for photomicrography. Either system can be adapted to photograph snow crystals, whose sizes commonly range from 0.5 to 5 millimeters. The microscope is most suitable for examining and recording the structural details of single crystals, for this requires a relatively large degree of magnification on the film. Less magnification is acceptable for routine recording of snow crystal types, so photomacrography is more often used for this purpose. Compact and portable equipment with short exposure times is best in the field, while the bulkier and less flexible camera-microscope combination finds its best use in the cold laboratory.

The two most famous publications on snow crystals, those by Bentley and by Nakaya, both present many beautiful photographs taken by photomicrography. Bentley used a microscope adapted for use with a large view camera in a bulky arrangement he built himself. His "cold laboratory" was the natural chill provided by Vermont winters. Nakaya used microscopes with standard photographic attachments in a modern scientific cold laboratory.

The photographs in this present snow crystal guide were all taken with 35 mm photomacrography equipment, both in the field and in a cold laboratory. Two types of 35 mm cameras were used. One was built into a compact unit which included camera body, lens, prism, light source, and batteries all in a single package. A fixed lens extension tube and snow crystal support assured good focus at a single plane but did not permit any adjustment. Lens position in respect to the film was chosen to give a 1:1 object-to-image ratio (crystal reproduced natural size on the negative). Mounted in a carrying case, this camera weighed only three pounds. It was readily carried in a rucksack and could be used very quickly in the field.

The other camera was a single-lens-reflex type mounted on a photomacrography stand with lens extension tubes and bellows, an adjustable snow crystal carrier, and a flexible light source. Focus, degree of magnification and illumination were all adjustable, while the reflex feature of the camera permitted visual examination of the specimen through the camera lens. Pictures were usually taken with an object-to-image ratio of about 1:4 (crystal magnified four times on the negative). This equipment was used in the cold laboratory.

There are several ways to illuminate snow crystals for photography (Fig. 8). Reflected illumination gives a white image against a black background (dark field illumination). Many of the pictures in this book were taken with reflected illuminations in both cameras. The portable snow camera provided low-oblique reflected illumination with additional diffused light. The photomacrography stand provided a choice of illumination methods. White crystals against a black background are often artistically attractive, and they have the further advantage of appearing as they are seen by casual examination with the naked eye. But reflected illumination does not reveal all the structural details of the crystals. Transmitted illumination passing through the crystals gives more information about internal structure, but does not bring out surface features. Transmitted illumination gives a white background in photographs (bright field illumination) and often a rather low contrast if the crystal is thin. Several pictures in this book were taken by direct transmitted illumination, most notably those of firn in Figure 45.

Maximum details of both surface and internal structure in snow crystals are revealed by low oblique transmitted illumination. Here the light source is off to one side of the camera axis, but light still passes through the specimen. Careful adjustment of the light source is necessary for optimum results. There is an option with this type

**Reflected
Illumination**

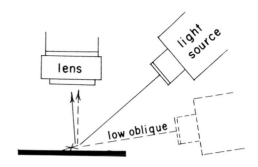

**Transmitted
Illumination**

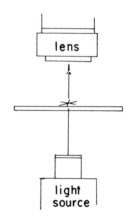

**Low Oblique
Transmitted
Illumination**

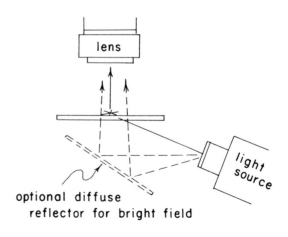

8. *Methods of illuminating snow crystals for photography.*

of lighting of providing either bright or dark field according to whether or not additional direct transmitted light is included. Most of the bright-field photographs in this book were taken on the photomacrography stand using diffuse light for low oblique transmitted illumination, with a small amount of diffuse light added by direct transmission.

The famous snow photographs of Bentley were all taken by transmitted illumination. Bentley achieved an apparent dark field by carefully cutting out each crystal from a duplicate negative and printing this so that the surrounding area was blackened. The cutting obviously can be a very tedious process for an intricate stellar crystal. Almost all of Nakaya's photographs were taken by low oblique transmitted illumination.

All of the photographs shown here were exposed on Eastman Plus-X film developed in Microdol-X. With 4X magnification on the negative, and oblique bright-field illumination, typical exposure on the photomacrography stand was ½ second at f.8. Light source was a PR-12 bulb operating on 6.3 volts. The portable snow camera used exposures of 2 seconds at f.12.7, with illumination provided by a 4.8 volt bulb operating on 4 size D flashlight cells in series.

The main problem in photographing snow crystals is to keep them from melting. A cold laboratory, which need be no more than a walk-in freezer, is the best place to work. A temperature around -10° C (14° F) is about right. A lower temperature is even better for the crystal but likely to be uncomfortable for the photographer. Very low temperatures raise the problem of frost forming on the camera optics every time the door is opened to admit warmer, moister air from outside. It is difficult but possible to work at temperatures just below the freezing point. Speed is essential to avoid melting, and it is necessary to take such precautions as chilling the slides that carry the crystals.

Strong illumination immediately melts snow crystals even in rather cold conditions. Interposing a layer of heat-absorbing glass will minimize this problem. The best solution is to avoid strong illumination, even at the expense of increased exposure times. A small flashlight bulb located 5-10 centimeters from the specimen works well. It is difficult to take crystal pictures in direct sunlight, especially with a black background which absorbs heat.

Rapid collection and handling of the crystals are essential. They metamorphose rapidly unless the temperature is very low. Evaporation is also a problem in a cold laboratory where the air tends to be dry because much of the moisture condenses as frost on the cooling coils. Moist air, especially from the photographer's breath, will deposit rime on cold crystals and even cause melting.

Some of the difficulties of working in a cold lab can be circumvented by preparing plastic replicas of snow crystals and then photographing the replicas. A chilled, dilute solution of polyvinyl formal in ethylene dichloride is dropped onto the crystal to be replicated; then this is set aside to dry and to allow the ice of the crystal to evaporate. Details of this technique are given in the article by V. J. Schaefer listed in the Bibliography. The plastic replica can be handled and preserved indefinitely at room temperature. Such replicas show a remarkable amount of detail if prepared carefully. The stellar crystals illustrated in Figure 65 are plastic replicas.

Discussion of the Photographs

In common usage, every kind of ice particle that falls from the sky is called a "snowflake." The term *snowflake* in fact has a more restricted meaning: it is an assemblage of *individual snow crystals* which have collided and remained fastened together during their fall through the atmosphere. Snowflakes are commonly made up of stellar crystals with intricately branching arms, because these tend most readily to remain hooked together. Snowflakes fall frequently when air temperatures near the ground are not far below the freezing point (the individual crystals may have originated at higher and colder levels), for the crystals cohere more readily at these temperatures. Snowflakes are observed less frequently at lower temperatures, but they do sometimes occur. Under favorable conditions they may achieve remarkable size. Flakes several centimeters across are occasionally seen in calm air near the freezing point. These are rapidly broken up or prevented from forming by turbulent wind.

The snowflake illustrated in Figure 9 is made up of large, branched stellar crystals which bear a few traces of rime particles (frozen water droplets).

Figure 10 is a single, large snowflake just as it fell on the collecting plate and was carried undisturbed to the camera. It is made up of many thin, clear stellar crystals and stellar fragments.

The common, familiar snow crystal is the stellar crystal. It will become obvious as this discussion evolves that there are many other kinds of snow crystals, but the six-rayed star is the one most frequently seen in nature and the one that first catches the observer's eye. These crystals are found in sizes ranging from tiny

9. *Stellar snow crystals combined to form a snowflake. 24X*

10. *A snowflake composed of many stellar crystals and fragments. 24X*

specks barely visible to the naked eye, all the way up to occasional specimens several millimeters in diameter. The very large stars seldom reach the ground intact, for they are readily broken apart by air currents. The wide variety of forms, ranging from delicate and intricate branches to simple rays or stubby arms, has given rise to the observation that "no two snow crystals are alike." Even if two were formed on an identical pattern, the chances are that they still would not be identical, for each can exhibit its own peculiar asymmetry. Stellar crystals perfectly regular in all details are definitely in the minority.

Figures 11 and 12 illustrate two typical stellar crystals greatly enlarged to the same scale under the oblique transmitted illumination which brings out maximum structural details. Figure 11 shows a very thin and rather delicate star with slender arms but surprisingly

11. Stellar snow crystal with a trace of rime. 30X

broad side branches. It exhibits an asymmetrical shape, and in addition the uneven but very light deposition of rime particles. The crystal in Figure 12 is thicker. (With this type of illumination, the degree of lighting contrast is roughly proportional to thickness.) The arms are wide and blunt, with each terminated in broad features resembling small plates (see types P1c and P2a in the classification chart, Figure 5). This latter crystal is completely free of rime.

The molecular structure of ice normally leads to hexagonal symmetry of snow crystals which grow in the plane of the crystallographic a-axes (stars, plates, and allied forms). Twelve-sided crystals (P4a and b) are not uncommon in nature, apparently in violation of this hexagonal symmetry. On close examination, these twelve-rayed stars turn out to be two normal hexagonal crystals superimposed one on the other. The two components can sometimes be separated

12. Irregular stellar crystal with broad arms. 27X

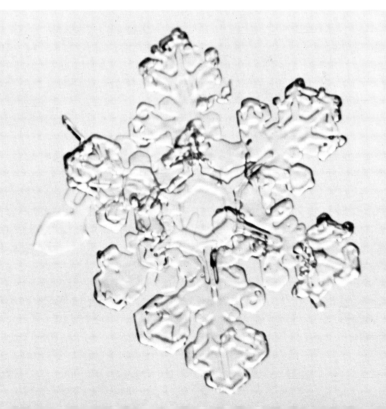

13. Six- and twelve-sided stellar crystals moderately coated by rime. 6X

by careful manipulation. Alignment of the two superimposed crystals is not always precise, leading to irregular lengths of the twelve arms and to angles other than 30° between the arms. The double-crystal character of these twelve-sided stars is obscured when they become coated with rime, as are the ones shown here in Figure 13. Normal hexagonal crystals and rimed stellar fragments are also visible.

Clear, delicate stars are likely to be very thin. The long, slender streak on the right-hand portion of Figure 14 is a large stellar crystal seen "edge-on." All of the crystals in this group are thin, rime-free stars or fragments of stars. When snow like this is deposited with little wind, it builds a very fluffy, feltlike layer with low density (60 kg/m^3 or less). The crystals interlock to give the snow a surprising amount of cohesion in spite of its fluffy nature. This cohesion causes the snow to form canopies on tree branches, fence posts, or other exposed objects. Under favorable conditions these canopies keep building up by the addition of interlocking crystals until they become much larger in diameter than the object on which they grew. As soon as the snow suffers destructive metamorphism and the

14. *Clusters of thin stellar crystals making up very fluffy new snow. 20X*

crystal branches break up, this cohesion from interlocking is lost. The snow becomes looser and less feltlike, the canopies fall, and the snow tends to slide off in little rivulets ("sluffs") from steep slopes. This often happens a day or two after the snowfall; the exact timing depends on temperature.

Figure 15 is a mixed collection of large stellar crystals and fragments. Some are thin, clear stars similar to those in Figure 14. Others are frosted by a coating of rime particles. The newly deposited snow on the surface from which this specimen was collected consists of about half of each. These two distinctly different crystals obviously experienced different growth histories in the clouds. They probably formed under similar temperatures and degrees of supersaturation, but one type passed through a supercooled cloud on its journey to earth while the other did not. It can be presumed that the rimed crystals formed above such a cloud, while the clear crystals formed below it.

Homogeneous snowfalls made up exclusively of just one single crystal type do not usually persist for long periods. Mixed snowfalls are more common. The observer needs to be alert for these mixtures and to watch for the shifting proportions of the different crystal types. There is a tendency for the eye to be caught by the larger and more striking crystals to the exclusion of the inconspicuous ones. This can cause misleading reports about the proportions of different types in a snowstorm.

The new-fallen snow shown in Figure 16 fell at —9° C during a light snowstorm with little wind. It is an example of the typical mixed snowfall the snow crystal observer is likely to collect for observation. All the crystals shown are lightly and rather evenly rimed. On first glance, the whole group might be dismissed as "stellar crystals." Closer inspection shows that two different types of stars are present. One is the small, simple star of fairly uniform size (P1d). The other is a large, intricate one (P1f), present here only in fragments. These two types undoubtedly originated at different levels in the atomsphere. There is at least one plate crystal plainly

15. *Large stellar crystals and fragments form new snow. Some of the crystals are rimed.* 24

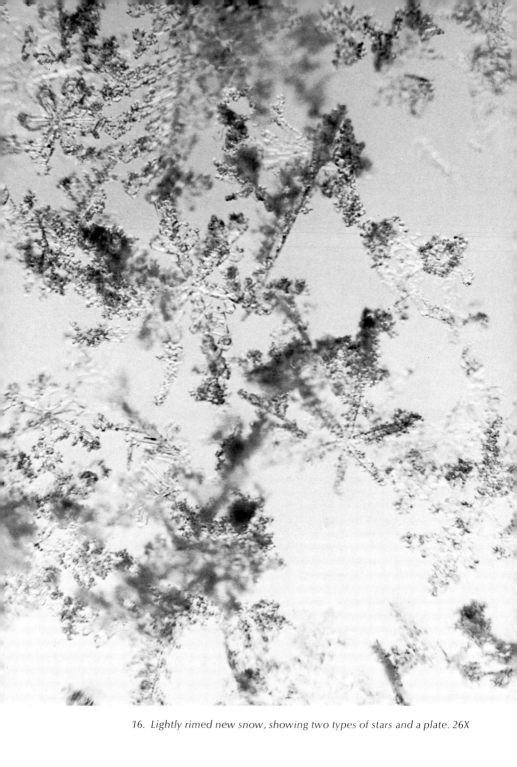

16. *Lightly rimed new snow, showing two types of stars and a plate. 26X*

17. A rimed spatial dendrite. 22X

visible. Several small rimed particles are also present; these seem to be smaller fragments of stars.

Snow like that in Figure 16 has a different character when deposited on the ground from that shown in Figures 9 and 14. It is less fluffy and has much less tendency to form canopies. Initial interlocking among crystals is limited; hence the snow is not so likely to go through a "sluffing" period a day or two after deposition. There is a greater tendency for this kind of snow to build up soft slab avalanches if it is deposited with any appreciable wind.

A distinct variation of stellar crystals occurs when the branches or arms are connected to each other randomly instead of building a pattern of hexagonal symmetry in a single plane. This produces a form called the "spatial dendrite" (F7b). Figure 17 illustrates a large spatial dendrite lightly and evenly coated with rime.

Nonstellar snow presents a distinctly different picture to the ob-

18. Columns, plates, and clusters of plates and sectors. 20X

server from that of the stellar kind. The snow in Figure 18 is typical of nonstellar crystals collected from the snow surface and spread out on a black background. It contains two or three small and poorly formed stars, but most of the crystals are columns (C1e, C1f), plates (P1a), and clusters of plates and sectors (P7a, S1). Compare the appearance of this snow with that in Figure 14. The details of these generally smaller crystals are more clearly seen by bright-field illumination, such as that of Figure 19. The latter photo shows several clearly delineated plates, both featureless and structured, as well as a prominent column. In addition, there are numerous clus-

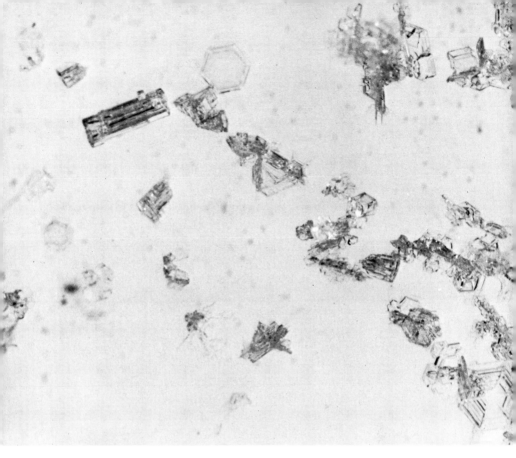

19. Columns, plates, and clusters of irregular crystals. 23X

ters of plates or sectors which form irregular particles difficult to classify. These types would fall most readily into the groups S1, S2, and S3. The specimens of both Figures 18 and 19 fell at temperatures in the vicinity of —9 to —12°C. Snow of this type tends to have a higher initial density than the fluffy snow composed of stellar crystals; values greater than 100-120 kg/m³ are common.

Clusters of sectors (sections of plates or stubby stars) are common at lower temperatures. They adopt a wide variety of forms, and this makes them difficult to classify. The crystals in Figure 20 illustrate some of the possible combinations. In the center is a prominent cluster of plain and mostly featureless sectors, rather large in size for this type. It would definitely be called type S1. (This discussion is adopting the term "sector" instead of the Magono-Lee term "side plane," which seems less descriptive.) Immediately adjacent to it is a combination of bullets, clearly C2a. Just beyond the latter combination and partly out of focus in the camera field is a more obscure form which appears to be composed of sectors resembling poorly shaped stellar branches, plus possibly a column. This specimen defies exact classification unless we take refuge in the "Miscellaneous" category (I4), but is most closely allied to S3. Also visible in the lower part of the photo are two more clusters which might be more truly called S3. The large, irregular, and featureless plate appears to be an isolated large sector unattached to any combination. The other particles are probably fragments of types already described.

The crystals in Figure 21 had already been on the ground for thirty-six hours when this photograph was taken. Equitemperature metamorphism has already started to alter the original shapes, but they can still be recognized. This alteration is noticeably stronger on the features resembling branches of stars or dendrites, which have already become rounded and smooth, while the columns and bullets are little changed. The more intricate shapes are less stable and hence tend to metamorphose more rapidly. Some of the bullets shown here belong in the category of bullets with plates (CP2a), although the plates are small and poorly formed. Several combinations or clusters are present, including one cluster of dendritic branches (all but one branch has its edge toward the camera). Most of the clusters come closest to category S3.

Even more complicated crystals are found in Figure 22. The large crystal is a capped column (column with plates, CP1a). One of the plates is much more heavily rimed than the other, presumably reflecting the fall orientation of the crystal, the more heavily rimed end being the lower side. Adjacent to this is a column, or bullet, with dendrite (CP2b) which is almost an exact duplicate of the type sketch for this form. Near the corner of the photo is a combination

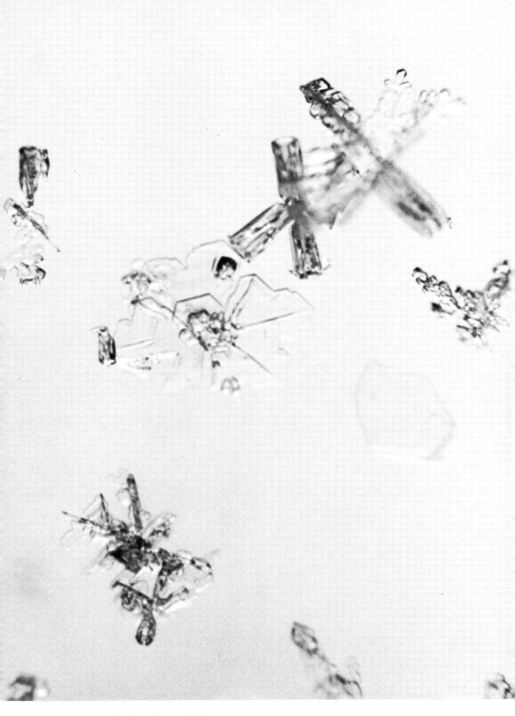

20. *Sector and bullet clusters. 30X*

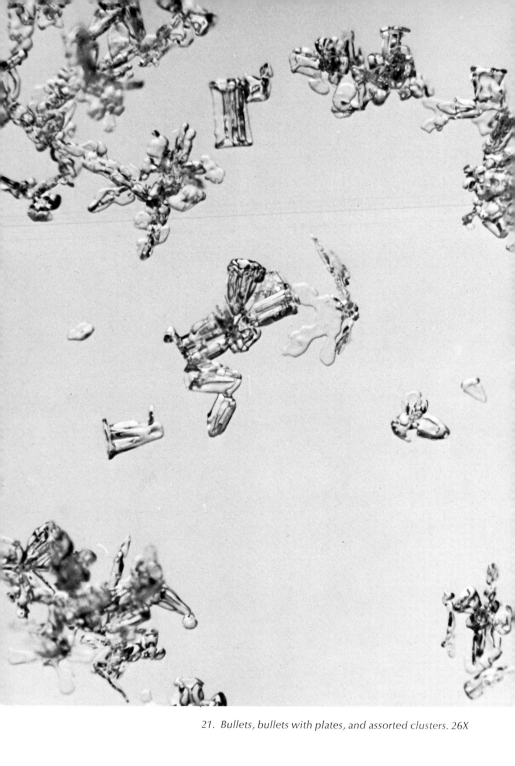

21. *Bullets, bullets with plates, and assorted clusters. 26X*

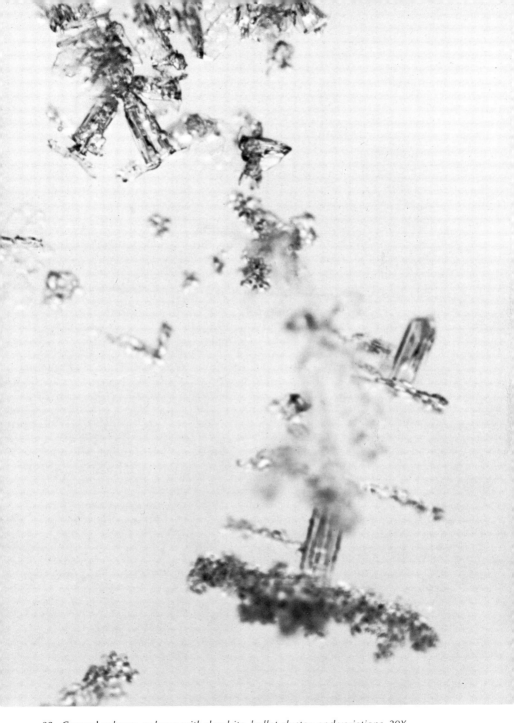

22. *Capped column, column with dendrite, bullet cluster, and variations. 30X*

of bullets with either plates or poorly formed dendrites (approximately CP2a). One of the bullets in this combination has a stepped appearance and a double set of plates at the base. It should be noted at this point that the capped column was called a "crystal," while the bullet form was called a "combination." The latter is not a crystal when taken in its entirety, for a crystal is any separate piece of ice with a common orientation of crystallographic structure. The a- and c-axes of the plates and column making up the capped column are all parallel; hence it is a true crystal. These axes are not parallel in the case of the bullet combinations (and clusters in general); hence the whole particle cannot correctly be called a crystal. It is an assemblage of several crystals.

Figure 23 is a beautiful example of a column with dendrites (CP1b). There is a small amount of rime on the dendrite branches.

Snow that falls at low temperatures often has a very fine, granular appearance. This is typical of snow in the polar regions, but also occurs at temperate latitudes in sufficiently cold weather. Such snow can be mistaken for fine-grained old snow on casual inspection, but close examination under a microscope shows the particles to be clearly patterned snow crystals rather than the rounded crystal grains produced by metamorphism. The snow in Figure 24 fell from a cold arctic air mass at temperate latitudes when the ground temperature was —21°C (—6°F). The crystals must have formed at even colder temperatures higher in the atmosphere. Their apparently granular character turns out to consist of very minute columns and plates. The columns are stubby, many of them almost as wide as they are long. Snow of this type readily packs into a firm layer with relatively high density of 150 kg/m³ or more. The crystals in Figure 24 are very similar to those in Figure 18 except that they are much smaller.

Needle crystals and their allied form, sheaths, frequently occur at ground temperatures near the freezing point. Some snowfalls consist entirely of these slender forms. Because riming conditions are so often found at the same temperatures that generate the needles,

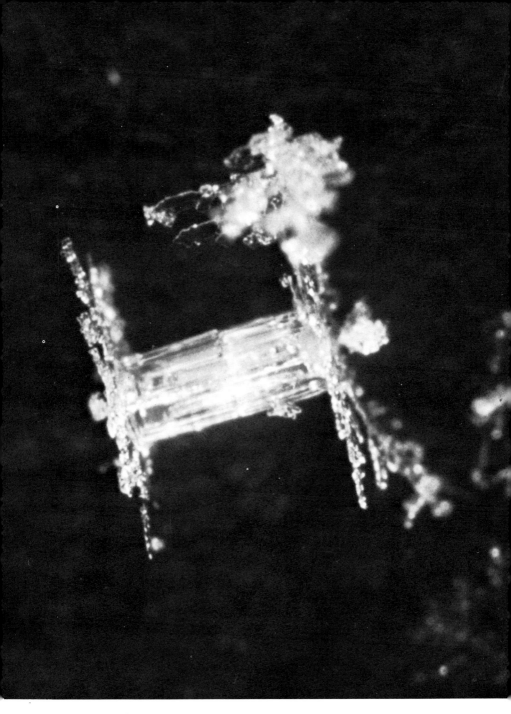

23. *Column with dendrites. 57X*

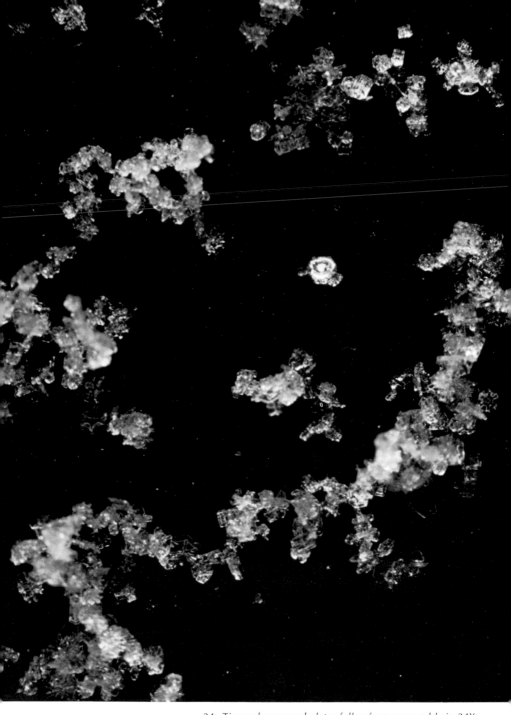

24. *Tiny columns and plates fallen from very cold air. 24X*

a high percentage of them are coated with rime. Their shape causes these crystals to pack closely together to form a dense, stiff snow layer. Newly deposited snow layers with densities as high as 330 kg/m³ have been observed. (This is a normal density for old, strongly metamorphosed snow.) Figure 25 shows a collection of lightly rimed sheaths taken from a snow layer with density of 170 kg/m³.

The needles in Figure 26 formed a new snow layer that produced vigorous avalanching. The high density and stiffness of snow layers composed entirely of needle crystals make them particularly favorable for slab avalanches. Prolonged snowstorms with a high percentage of needles and strong winds are one of the surest warning signs of snow avalanche danger. Riming on the needles seems to contribute to this behavior. Note that the needles in Figure 26 are either very heavily or very lightly rimed. From this it can be deduced that they originated at two different levels in the storm clouds, one but not the other having fallen through a substantial supercooled region.

Figure 27 is another example of rimed needles. These fell at —2°C. In this case the riming is so heavy that the needles are almost completely obscured. This snow layer also produced vigorous avalanching.

Still another type of snow associated with avalanche formation is the stiff, wind-drifted slab. This type is produced by strong wind action which breaks up stellar crystals into many fragments. Often these fragments are coated with rime which contributes to the high density and increased cohesion of the deposited snow. Particles of this type are shown in Figure 28. Some can be recognized on close inspection as stellar fragments, while most are either so broken or so coated with rime that their original form cannot be distinguished.

The formation of rime on snow crystals has already been mentioned and illustrated several times. Rimed snow is common in those climates which produce abundant supercooled water clouds. Relatively warm conditions (0 to —15°C at the levels of snow formation in the atmosphere), strong convection, and the rapid lifting

25. *Lightly rimed sheath crystals. 26X*

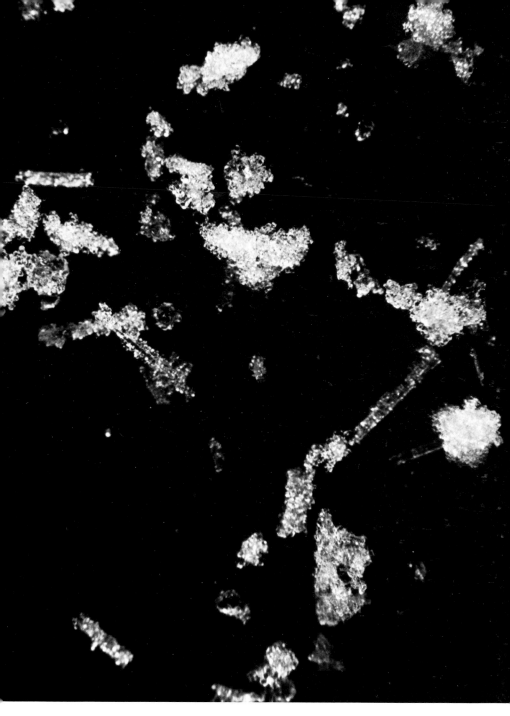

26. *Needle crystals, some heavily rimed. 24X*

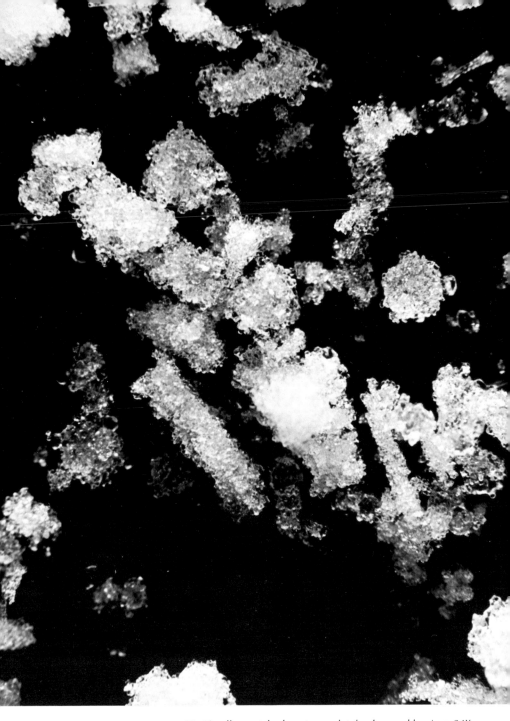

27. *Needle crystals almost completely obscured by rime. 24X*

28. *Stellar fragments produced by wind-drifting. Some are coated with rime. 24X*

of moist air by steep mountain ranges all favor rime formation on snow crystals. In some areas where all these conditions prevail, rime normally is found on 90 per cent or more of the falling snow. Cold, dry, continental climates, on the other hand, more commonly produce rime-free crystals. The degree of riming on snow, and the mixture of differently rimed crystals, can furnish information about the patterns of temperature and cloud layering in the atomsphere. In Figure 29 there are two heavily rimed stars (R2b), one very lightly rimed smaller star (R1d), and a single rime-free plate (P1a). Comparing these with the Magono-Lee diagram leads to the conclusion that the stars originated around —15°C and then fell through a supercooled cloud, the lightly rimed crystal having started at a lower elevation and fallen a shorter distance within this cloud. The plate

29. *Various degrees of riming on stellar crystals and a plate. 32X*

must have formed in an ice cloud somewhere between —12° and —18°C. The higher temperature is most likely, for the plate did not accumulate any rime and hence must have originated below the supercooled water cloud where the air was warmer.

The peculiarities of rime deposition on falling snow crystals are determined by the aerodynamic properties of the individual crystals as well as by the character of the various air layers through which they fall. Figure 30 shows two large stellar crystals collected side by side in the same snowfall. One is evenly but lightly coated with rime particles. On the other, more skeletonlike, form, rime has been heavily deposited at the branch tips but hardly at all in the center. One branch has broken off, very probably under the burden of its unbalanced load of rime.

30. *Differences in riming on large stellar crystals. 21X*

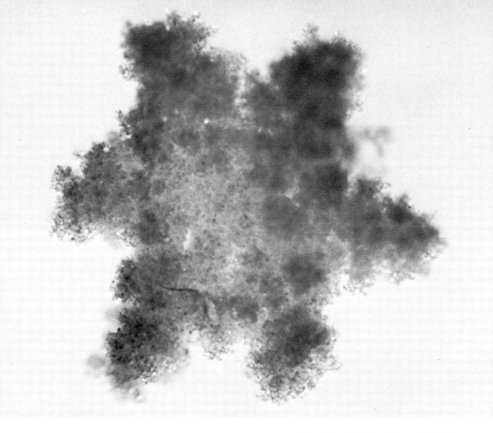

31. Heavily rimed large star. 25X

Figure 31 shows a large stellar crystal (R4a) so obscured by rime that only the crude hexagonal shape remains to suggest that a star formed the original framework on which the rime was deposited. This is the last stage prior to the formation of *graupel* (R4b and c), which is the name given to snow crystals completely enveloped in rime. Graupel is often called "pellet snow" by skiers, or "soft hail" by meteorologists. The large, dark lumps in Figure 32 are pellets of graupel, mixed with assorted rimed stellar fragments and other obscure crystal forms. This combination is a common one. In certain climates an observer will see this kind of snowfall more frequently than he will the kind of snow shown in Figures 9 or 14.

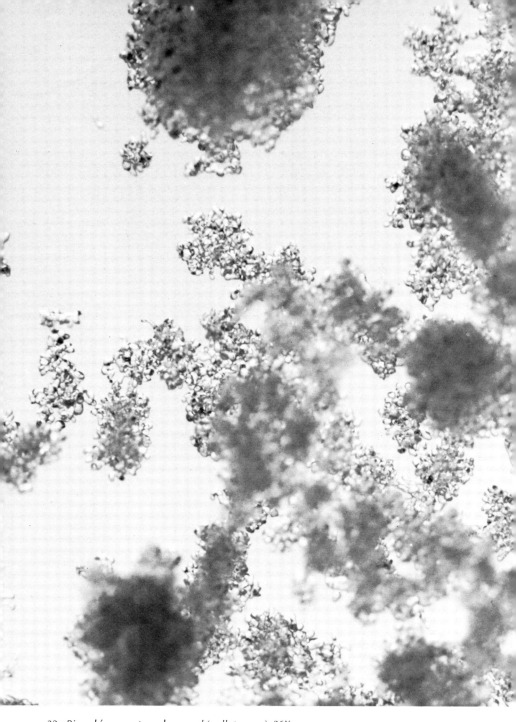

32. *Rimed fragments and graupel (pellet snow). 26X*

A fully developed example of graupel, the kind that earns it the name "pellet snow," appears in Figure 33. Deposited snow composed entirely of such pellets has some peculiar properties. At warmer temperatures it forms a compact, stiff layer that provides a skiing surface much like that of a packed ski run. If the layer is thick enough and has been deposited during high wind, it can cause serious danger from slab avalanches. At lower temperatures and with light wind, graupel tends to have little cohesion. Streams of pellets roll and bounce down the steeper slopes, forming trickles and even substantial currents of flowing snow which behave like miniature and continuous loose-snow avalanches. Graupel also persists for a long time within the snow cover. The pelletlike character of the snow particles can sometimes still be recognized months after deposition. The rounded lumps of graupel, composed of tiny, rounded particles of rime, have already adopted the stable shape which equitemperature metamorphism tries to produce; hence there is little room for further metamorphism. Compare the new-fallen pellets in Figure 33 with those in Figure 34, which are a month old.

Once crystals have been deposited in the snow cover, metamorphism begins to alter their form. The further this progresses, the more difficult it is to recognize the original types. The crystals with the most intricate shapes, such as the stellar and dendritic types, change most rapidly at a given temperature, for their shape is the most unstable. The initial effect of metamorphism in the absence of a temperature gradient is the rounding and smoothing of structural details. The sharp points and edges become blunt, and the intricate surface details begin to vanish. Figure 35 shows large, rime-free stars and stellar fragments in this first stage of such metamorphism. (Examples for this discussion of metamorphosing crystals have all been chosen with oblique transmitted illumination for maximum detail.) These crystals are forty hours old; compare them with the newly fallen stellar crystals in Figures 9 and 11. Another twenty-four hours later, snow from the same layer has progressed to the stage shown

33. *Fully developed graupel, or pellet snow. 24X*

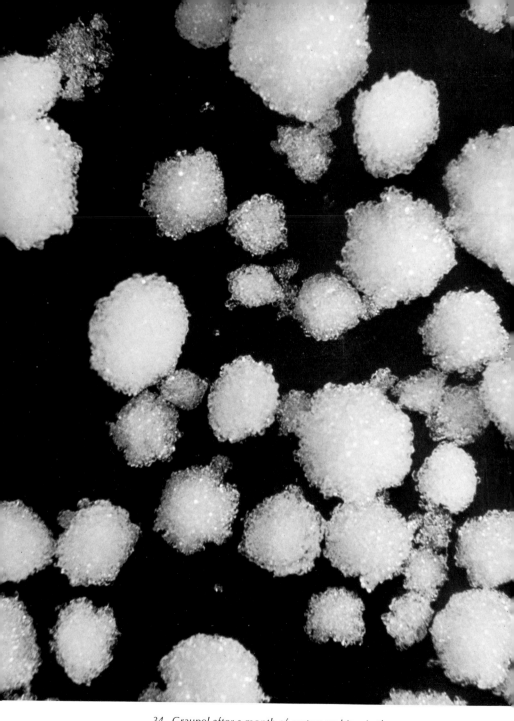

34. *Graupel after a month of metamorphism in the snow cover. 7X*

35. *Stellar crystals in the first stages of equitemperature metamorphism. 26X*

36. The same snow layer represented in Figure 35, after another 36 hours of meta-morphism. 26X

in Figure 36. The large stellar crystals have been broken up, and the remaining fragments have lost more structural details. By this time the original crystal forms can be recognized only with difficulty, and then in many cases only if the previous stages of metamorphism are known.

Other crystal types experience a similar progression. Those in Figure 37 were originally a mixture of lightly rimed stellar fragments (I3b) and rime-free clusters of side planes, or sectors (S1). Meta-morphism has blurred original shapes and started to break up the clusters. This sample is only twenty-four hours old. Changes have been rapid at the temperature of —2°C.

37. *Early equitemperature metamorphism of mixed crystal types, including sector clusters. 26X*

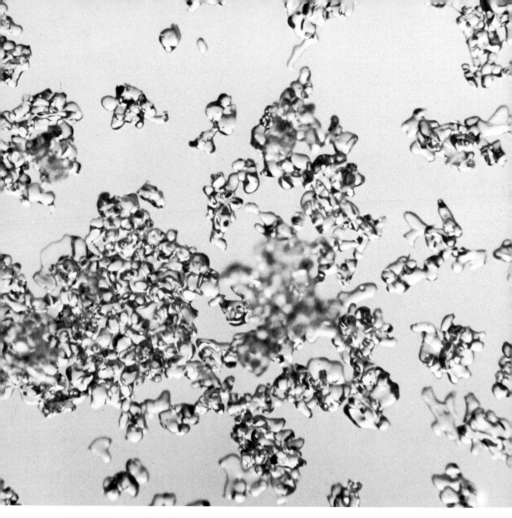

38. Stellar crystals which have lost almost all their identity through equitemperature metamorphism. 26X

As destruction of the original crystals proceeds, they become completely unrecognizable, although occasional rounded fragments may suggest the last remnants of stellar branches (Fig. 38). Other crystal types lose their identity more quickly, except for graupel, as previously mentioned. The particles in Figure 38 are still very small, but the continued metamorphic process causes the larger ones to

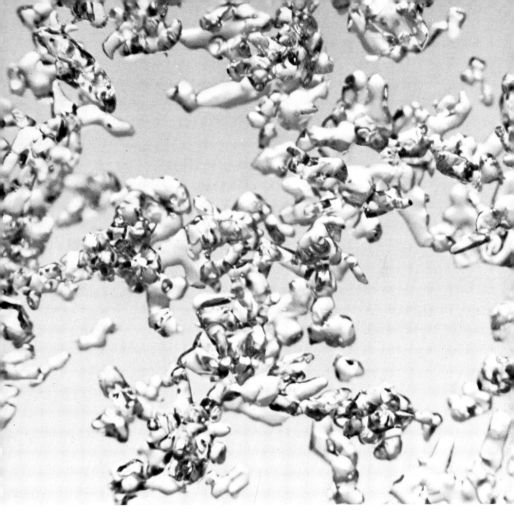

39. Advanced equitemperature, or destructive, metamorphism. 26X

grow at the expense of the smaller ones. The number of individual snow crystals becomes smaller as their size becomes larger, even though they may still be far from reaching the completely rounded grains of old snow. This more advanced stage is shown in Figure 39, a sample of strongly metamorphosed snow from a layer that has reached the density of 250 kg/m³.

Near the freezing point, metamorphism proceeds very rapidly. The large, round grains of old snow may begin to appear even before destruction of the original crystals has been completed. An example of this is seen in Figure 40, where large ice spheres have formed at —0.5°C in snow that is only twenty-four hours old. This phenomenon seems to be limited to temperatures at or just below the freezing point. At lower temperatures, a slower and more uniform metamorphism prevails.

Modern skiers have widely used the term "powder snow" to mean any loose, low-density snow that provides good skiing. It is usually applied to fluffy, new-fallen snow. "Powder snow" in fact has an older and more restricted meaning. This is snow that has suffered partial metamorphism so that it loses its original fluffy character and becomes more powderlike. The interlocking of stellar crystals is destroyed, and the snow loses some of its cohesion. Especially on shaded slopes that remain cold, this development of powder snow is furthered by temperature-gradient metamorphism. Near the surface this latter process is only partially effective, for the temperature gradients fluctuate daily in magnitude and even in direction. The result is snow that partially experiences both types of metamorphism, sometimes alternately. Over several days of clear, cold weather at higher altitudes, this process can reduce even stiff and compact surface layers to a loose powder. It improves skiing remarkably. Some of this true powder snow is shown in Figure 41. Mixed with the rounded and broken particles stemming from destruction of the original stellar forms are many angular, faceted crystals which indicate the first effects of a temperature gradient. This snow is seven days old.

The product of equitemperature metamorphism is by definition *old snow*. As produced in the normal winter snow cover at subfreezing temperature, this old snow has a very fine-grained, granular appearance. It may be cohesive enough to make shoveling difficult and exhibit considerable mechanical strength. The ultimate strength depends in large measure on how much compaction it has experi-

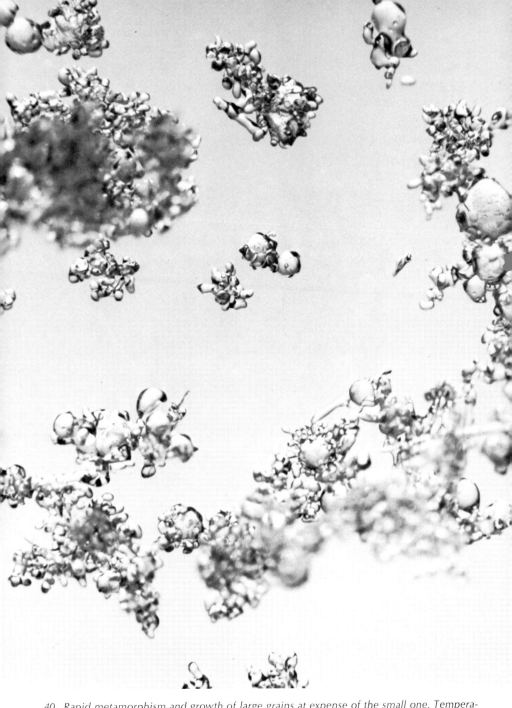

40. *Rapid metamorphism and growth of large grains at expense of the small one. Temperature is near the freezing point. 26X*

41. Fine-grained old snow reduced to "powder snow" by temperature-gradient metamorphism. 26X

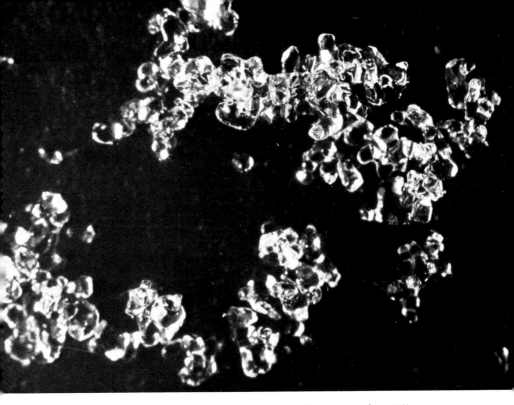

42. Fine-grained old snow. Equitemperature metamorphism is complete. 26X

enced under the load of successive snowfalls. Close examination of such snow (Fig. 42) reveals small, rounded crystal grains which often are uniform in size within a given snow layer.

Figures 43 and 44 give additional examples of old snow under oblique transmitted light where the details of the grains are more prominent. Both are from snow layers with a density of 390 kg/m³. The snow of Figure 43 is about three weeks old, that of Figure 44 about six weeks old. The younger snow was deposited in midwinter in a deep snow cover and has actually experienced more metamorphism and grain-size increase than the older snow. The primary difference is due to temperature and temperature gradient. The snow six weeks old was part of a shallow snow cover early in the winter which was exposed to lower temperatures and appreciable

43. *Fine-grained old snow, 3 weeks old. 26X*

44. *Fine-grained old snow, 6 weeks old. 26X*

temperature gradients. The latter delayed its progression toward old snow. (It should be remembered that "old snow" by definition is a physical state, not a given chronological age.)

Snow that persists into the summer usually does so at the freezing point in temperate climates. Liquid water is present throughout the snow layers, and the temperature, 0°C, is everywhere the same. Metamorphism can continue under favorable conditions for as much as several months. The result is further rounding of the grains and growth of the large ones at expense of the smaller. Sintering continues to augment the crystal bonds. The product is the fully developed firn shown in Figure 45a, a sample about six months old collected from the accumulation zone of a temperate glacier. The density is 610 kg/m³. The crystal grains are thoroughly rounded, some almost spherical, and bonds are well developed. Some of the latter have probably been produced by refreezing of interstitial meltwater in the spring. Figures 45b, c, and d are samples of older firn taken from a deep pit at the same site as the sample in Figure 45a. (The scale divisions in all four photos are millimeters.) The firn in Figure 45b is about eighteen months old. The individual grains differ little from those of the younger firn in Figure 45a, but bonds formed by refreezing of meltwater during the previous winter have multiplied, giving some angular facets when the sample is broken apart to be photographed. Density is 640 kg/m³. Figures 45c and d show firn samples three and six years old, with respective densities of 700 and 840 kg/m³. Even at the age of three years, the individual grains have not grown appreciably, but the effects of compaction and refreezing have built up substantial bonds that make it difficult to break apart the grains. Many fractured areas are now visible in this disaggregated sample. By the age of six years, burial under 15 meters of firn has severely compressed the original snow. Individual grains can no longer readily be broken from the compact mass, and only irregular fragments unrelated to crystal boundaries can be separated. This is no longer snow or firn; it is the initial stage of glacier ice.

The snow in Figure 46 bears a superficial resemblance to firn, especially that of Figure 45c. Nevertheless, it has quite a different origin and age. This snow is about three months old and has a density of 350 kg/m³. It has far less mechanical strength than three-year-old firn. This is old winter snow which has experienced a substantial amount of temperature-gradient metamorphism and consequent crystal growth. As the layer was buried deeper in the snow cover by subsequent accumulation, the temperature gradient diminished and the process reverted to destructive metamorphism. This erased many of the features of sublimation, but left the large, irregular crystals resembling old firn. The considerable differences in age, density, and strength provide ready clues to distinguish this from firn, as well as the fact that firn will seldom be found in a transient winter snow cover.

Depth hoar is the endproduct of temperature-gradient metamorphism. This snow has been completely regenerated into entirely new structures that bear no visible relation to precipitated snow. A stepped or layered character of the crystal faces is the common identifying feature of depth hoar. The crystals may take many shapes; cups and scrolls are characteristic, but these actually are a minority. Hexagonal patterns are common, but square corners are often seen. Many crystals, though not true cups, are hollow. Figures 47 and 48 show typical assortments of depth hoar crystals. These crystals sometimes reach a large size—up to 10 mm or more in the normal confines of a winter snow cover. Many of the common shapes of depth hoar are visible in these two photographs, where a sample of snow in each case has been broken apart to display the individual crystals. Figure 49, on the other hand, shows an undisturbed sample of depth hoar with the crystals arranged in their natural relationship. Two things are obvious: the somewhat orderly disposition of the crystals, and the large amount of empty space among them. This high porosity of depth hoar layers, together with few and poorly formed intercrystalline bonds, causes the very weak mechanical structure of this snow type.

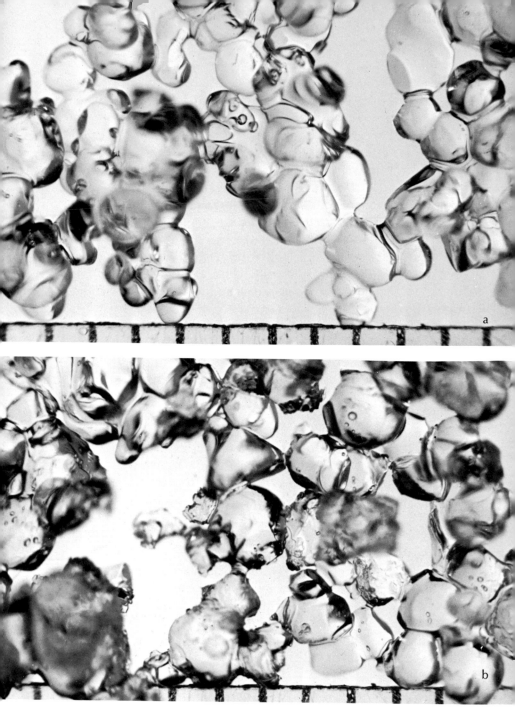

45. *The sequence of firnification in snow: (a) 6 months; (b) 1¹/₂ years;*

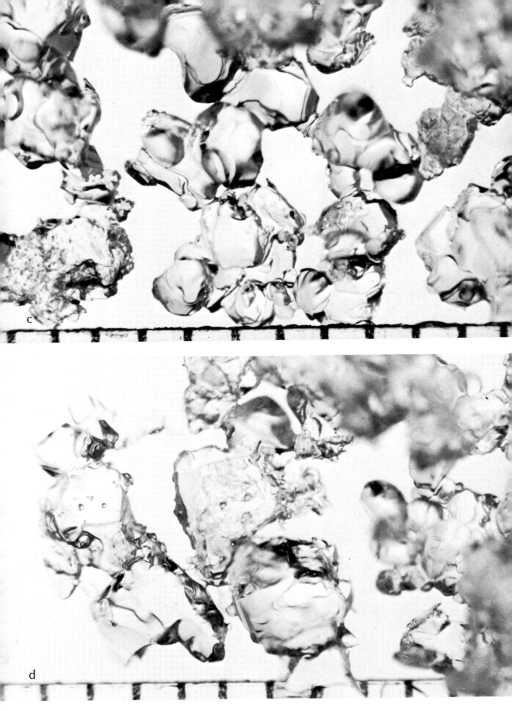

(c) 3 years; (d) 6 years. All 12X

46. *Snow 3 months old which has been exposed to temperature gradients, then has reverted to equitemperature metamorphism. 28X*

47. *Typical crystals of mature depth hoar. 24X*

48. Typical crystals of mature depth hoar. 24X

49. *Undisturbed sample of depth hoar showing natural arrangement of crystals. 24X*

The initial kind of snow, its temperature, and especially its temperature gradient, all determine the speed with which depth hoar will form and the ultimate size the crystals will achieve. Very strong gradients at warm temperatures cause depth hoar to form very rapidly. The process usually starts in relatively low-density snow; it can proceed only with difficulty when the initial density exceeds 400 kg/m³. The snow in Figure 50 is typical of new snow in which the effects of temperature gradients start to work. It is twenty-four hours old, is composed of stellar fragments (some rimed), has experienced a little equitemperature metamorphism, and has a density of 150 kg/m³. This snow was placed in a cold box under controlled conditions which produced a temperature gradient of 2°C/cm, larger than is usually found in nature. Large, mature crystals of depth hoar formed within six days. The complete sequence of depth hoar formation is shown in Figure 51, with stages after application of the temperature gradient as follows:

Figure 51a	23 hours
Figure 51b	48 hours
Figure 51c	71 hours
Figure 51d	95 hours
Figure 51e	120 hours
Figure 51f	142 hours

At twenty-three hours, the first stages of temperature-gradient metamorphism have appeared (CM1). By forty-eight hours the first layered crystals of depth hoar are showing (CM2). After seventy-one hours the partial stage is well advanced and recognizable cup crystals can be seen. By ninety-five hours metamorphism is complete (CM3) and depth hoar has been formed. The process still continues and the crystals keep growing, ultimately reaching the very large cups shown in Figure 51f. These photographs are all reproduced to the same scale, so that the magnitude of crystal growth can be clearly seen.

50. *Stellar fragments 24 hours old. This is the snow used to start the sequence of depth hoar formation (see text) shown in Figure 51. 29X*

51. The stages in the formation of depth hoar beginning with the snow pictured in
Figure 50. Temperature gradient is 2°C/cm. (a) 23 hours; (b) 48 hours; (c) 71 hours;

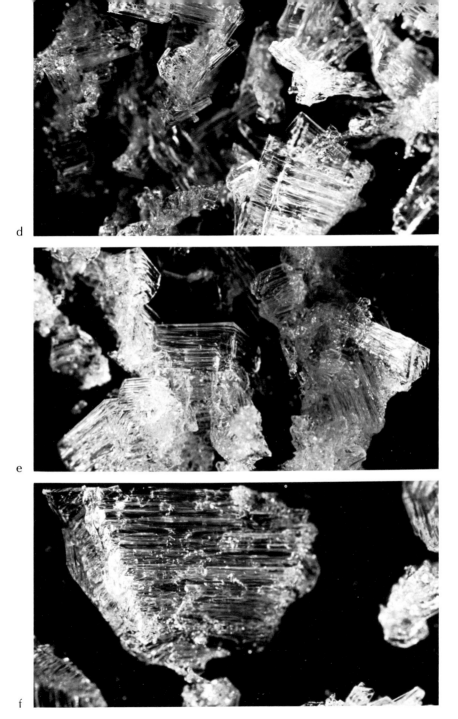

d

e

f

(d) 95 hours; (e) 120 hours; (f) 142 hours. All 14X

52. Depth hoar crystals slightly altered by equitemperature metamorphism. 20X

The large, open depth hoar crystals ("cups" or other hollow forms) tend to form more readily when a temperature gradient is imposed on coarse-grained, older snow. Smaller, more compact crystals are the rule if the recrystallization begins in fine-grained or newer snow. In either case, there is a large loss of mechanical strength.

Once the temperature gradient is removed, equitemperature metamorphism resumes and the crystals start to lose their layered structure. Figure 52 depicts large, mature depth hoar crystals which have started to decay in this fashion. Ultimately they will reach the condition shown in Figure 53, where the original depth hoar is barely recognizable. Snow like this regains much of its strength

53. *Depth hoar crystals profoundly altered by equitemperature metamorphism. 30X*

over a period of two to three months, but always remains weaker mechanically than snow that has not been subjected to a strong temperature gradient.

There are several nonprecipitated forms of ice crystals frequently encountered by the winter snow observer. These do not fall from the sky or develop by metamorphism from crystals that do.

The most familiar nonprecipitated form is *rime,* whose mode of formation from supercooled water droplets has already been described. The rime in Figure 54 has been built up on a dead snag located at the top of a steep ridge. The wind blew upslope from the left, producing the tilted rime feathers which grew *into* the prevailing storm wind. Small, individual rime feathers are shown in Figure 55. These were collected from an exposed wire after a storm that was accompanied by a supercooled cloud at the ground level. These feathers are randomly oriented in the camera field, but each shows the direction it grew into the wind. Each rime feather starts at a small point and widens as it grows into the wind, developing a triangular and sometimes branching shape with the base at the point of origin. Under certain conditions which are not clearly understood, the rime particles accumulate in slender needles, or "rime spikes," as the wind-driven supercooled droplets pile up one on top of another. The spikes in Figure 56 are growing on the windward side of a ski pole that stood outside during a foggy winter night. Under the photomacrography camera (Fig. 57), these spikes resemble a giant version of lightly rimed needle crystals, but in this case there is no such crystal to form the core of these slender shapes.

High mountain peaks in maritime climates are exposed almost continually to high winds and supercooled clouds each winter. Very heavy accumulations of rime can form which completely blanket the rocks on even vertical faces. Wind-driven snow mixed with rime is also cemented onto these peaks. The result is a characteristic rough, channeled, and sometimes feathered deposit like that shown in Figure 58. The contrast between these rime deposits and the

54. *Rime feathers formed on a timberline snag.*

55. *Individual rime feathers collected from an exposed wire. 8X*

56. Rime spikes formed on a ski pole. 0.9X

57. *Individual rime spikes from the ski pole in Figure 56. 7X*

58. *Mt. Olympus, Washington, in February, festooned with a winter coat of rime and rime-cemented snow.*

smooth surface of normal, wind-drifted snow is readily apparent. These are the peaks of Mt. Olympus in Western Washington in mid-February.

As spring takes the place of winter, increasing intensity of sunlight causes more rapid melt of snow. This tendency to melt persists even when the air temperature is below freezing. Especially at high altitudes, cold air, radiation cooling, and evaporation all work together to chill the snow surface while sunlight passes through to cause melt in the subsurface snow layers. If these factors are balanced in just the right fashion, the meltwater from snow is refrozen

59. *Firnspiegel, or firn mirror, a thin sheet of clear ice formed at the snow surface in certain weather conditions (see text).*

60. *A section of firn mirror under the photomacrography camera. 5X*

at the surface to form a thin layer of clear ice. This ice then acts like the glass in a greenhouse to trap heat underneath and promote further snowmelt. The result is a surface of thin ice bridging hollows in the melting snow surface (Fig. 59). This ice has a highly reflective surface which has led to its name of *Firnspiegel,* a German word meaning "firn mirror." From a distance a whole slope may catch and reflect sunlight like a mirror to give the phenomenon of "glacier fire." This ice mirror is a thin, often perforated layer full of air bubbles. It is made up of many individual ice crystals. The crystal boundaries are very faintly visible in Figure 60, a photomacrograph of *Firnspiegel.*

95

Hoarfrost is produced by deposition of water vapor from the atmosphere under circumstances explained earlier. The result is a coating of surface hoar crystals like small, flat leaves or plates. They make a fuzzy blanket on the snow surface (Fig. 61; the scale is in centimeters) which glitters in sunlight. A thick carpet of surface hoar brings delight to the heart of skiers as well as beauty to the landscape. It makes an extremely fast skiing surface in which the passage of skis is accompanied by a characteristic rustling sound.

61. Surface hoar, a layer of sparkling crystal on the snow surface formed from water vapor in the air.

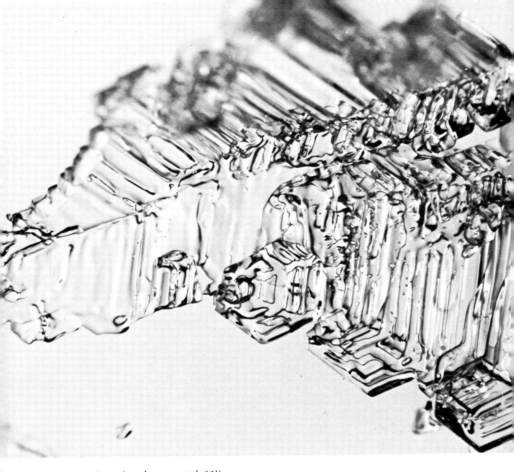

62. A single surface hoar crystal. 11X

Surface hoar also provides an excellent lubricating layer for slab avalanches. Close examination of an individual surface hoar crystal (Fig. 62) shows it to have the layered structure peculiar to all hoar crystals, including depth hoar. Like rime feathers, these crystals start at points and spread out as they grow longer. The narrow end of the crystal in Figure 62 is the base.

Hoar crystals may reach considerable size if growth conditions persist in a quiet environment. A long spell of clear, cold weather will sometimes generate hoar crystals several centimeters in diame-

63. Hoar frost crystal from a deep-freeze chest. This cup shape is also commonly seen in depth hoar. 29X

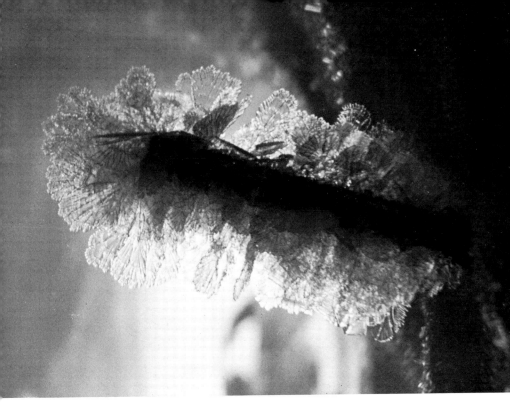

64. Crevasse hoar growing on a wooden stake in an ice tunnel wall.

ter on rocks, vegetation, or snow surfaces adjacent to a good supply of moist air, such as an open stream. Deep-freeze cabinets or cold storage lockers will also generate large hoarfrost crystals if left undisturbed for long periods. The crystal in Figure 63 came from the lid of a cold chest in a laboratory; it has a cup shape similar to that often seen in depth hoar. Enclosed ice spaces, such as ice tunnels or glacier crevasses, are also breeding places for hoar crystals whenever a temperature gradient persists along the walls. Very large hoar crystals, sometimes called crevasse hoar, can form if the growth remains undisturbed for months or years. The large, delicate flakes in Figure 64 are crevasse hoar which grew on a 2x2 cm wooden peg in the wall of an ice tunnel. This growth took about eight months to develop.

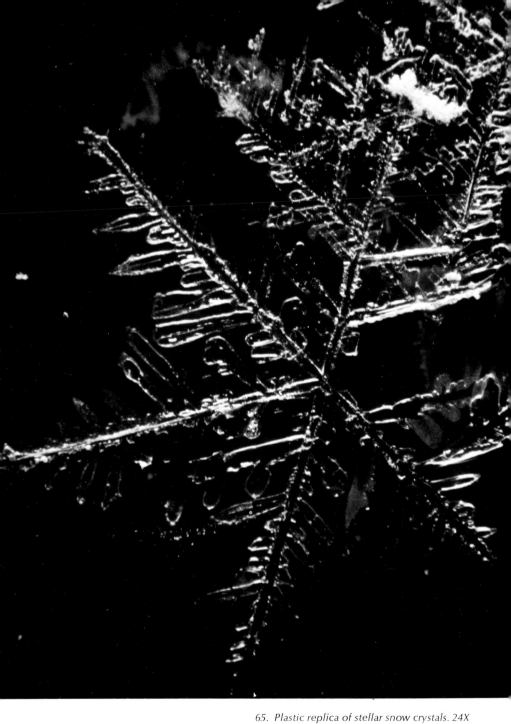

65. *Plastic replica of stellar snow crystals. 24X*

Bibliography

Bentley, W. A., and W. J. Humphreys. *Snow Crystals.* New York: McGraw-Hill Book Co., 1931; paperback ed., Dover T287, 1962.

Knight, Charles A. *The Freezing of Supercooled Liquids.* (Van Nostrand Momentum Book No. 14.) Princeton, N.J.: D. Van Nostrand Co., 1967.

Magono, C., and C. W. Lee. "Meteorological Classification of Natural Snow Crystals," *Journal of the Faculty of Science, Hokkaido University,* Ser. VII (Geophysics), II, No. 4 (November, 1966), 321-35.

Mason, B. J. *Physics of Clouds.* Oxford: Clarendon Press, 1957.

Nakaya, U. *Snow Crystals: Natural and Artificial.* Cambridge, Mass.: Harvard University Press, 1954.

Schaefer, V. J. "Preparation of Permanent Replicas of Snow, Frost and Ice," *Weatherwise,* XVII, No. 6 (December, 1964), 278-87.

Wood, Elizabeth A. *Crystals and Light.* (Van Nostrand Momentum Book No. 5.) Princeton, N.J.: D. Van Nostrand Co., 1964.